I0472569

The Secrets in The Woods

Part 2

When Raven answer the phone, while breathing uncontrollably, he just listens for a few seconds and hung the phone up. The expression on his face looks like he heard, that there were others in the woods also.

Since the woods were so dense, his tracking ability came to a halt, and he camps out at the edge of the dam, awaiting on a movement.

The Investigator, Agent Quinney, went back to the town, you gather more Information about the scientist, Dr. Raven.

So, he calls his agents in, and ask them about what they think is going on, in the woods. One of them said, 'Sir, the woods have too many secrets in it."

He said," how can this type of horrible situation go on, without any worries from the community, with a confused look on his face.

Agent Quinney said," I want you two, to monitor the activities they go on outside of the woods, while the rest of us, question the town folk.

Stephen and Cartina were still in the cave, they finally got back to normal, and they were hungry.

Stephen knew that they were being hunted, but the rush from his stomach was too much to control and so he said to her," baby, I will be back, I am going to get some food from the creek bed, do not leave this cave,

with a sincere look on his face, and she said to him," whatever you do, I will come with you, I will never leave your side while we are in here.

While it was early that morning, the sun has just come up, the ground was still wet, and the woods were alive with many noises.

Raven got up and started the hunt back for them. He was looking around every corner, and while the dense fog rose from his mouth.

Then It hit him, where they were, and he took off with haste. While Stephen was bending down at the creek getting some trust liquid for them to drink, Cartina was in her thoughts.

She remembered the time they were going on a horse ride, through the plain.

How Stephen was looking with his tight blue jeans on and they sexy body that she ever saw. As he was walking toward her, he put his hand out and said, 'grab my hand baby, and he put her a top of the horseback.

While still in a trance, Stephen was saying, "baby, 'baby, then she came out of the moments and said," sorry baby, I just had a moment."

Now Stephen is aware that Raven is coming after them, but he is trying to think, while the hungry was taking over.

Suddenly while standing by the creek bank, here comes a dead fish, floating down the stream.

The smell was horrible, but the illusion of meats that had been unjust, and Stephen Jump into the water and pick it up.

Without any hesitation, they consume it like wolves on prey. As they were approaching the cave, there was a sound inside, like an animal looking for food, so they easy back and bent down by a tree stump, hold the steam within, while the frost was freezing their bone.

An about that time, they saw a gun, sitting at the entrance of the cave, and then arose Raven, he facial look like a killer.

So, they just sit there, in the cold, while he moved from the entrance. Stephen was thinking, about that time, they were playing hide go see, with their children, but the thought did not last long, because Cartina said," he has gone back in, so let go, and they did.

The Agents was trying to figure what is more to this situation that meets the eye.

So, he looks up the Scientist name, started to figure out more about him.

One something he notices, that he is from this small town here, and his father was well known Minister.

His father led a group of followers and got release from the church for misbehavior.

About that time, he came across a picture, of some members of a certain organization, and there was His father, himself, and three others, but he could not put this together with the point at hand.

So, he calls the other agents, and ask them about ant activities they may have occurred, and they said," it quite here boss" and they hung up the phone.

As he was leaving the office in the small town, he noticed that most of the peoples there was looking at him strange, like they knew was is going on out there.

So, he got in his car, with a copy of the picture in his hand, and headed to the Sheriff station.

After his arrival, he saw some strange symbols, that seem to be everywhere, but he did not pay it attention.

10

While Raven was waiting for them to get back, Catrina said to Stephen, 'we can't keep this up, and expect to find our way out", with a frightening voice,

and he said to here," do you remember what I told you the other day," in order to have killed a lion, you must know how to out swift him, so here is our plan.

While at the sheriff office, Agent Quinney is looking for some answer quickly, if there any chance of finding out, is anyone still alive.

So, the sheriff came in and said," are you having luck on what in the hell is going on in those woods.

I want to tell you a story about time past.
My deputy will not go in there, because
of the secrets. This country has lost
plenty of peoples in time past.

You see, the loss we just suffer, it like
someone know when someone enters
the woods.

So, the Agent asks, who jurisdiction it is
the monitor the wood out there" and he
said the Rangers, in which are probably
dead, and the city police, but they don't
even patrol the area with a concerned
look on his face.

So, the agents pull out a picture of the Scientist, and his father on the picture, and ask the Sheriff, can you tell me about any of these peoples her.

While looking at the Sheriff and he said," when I was young, his father was the town minister, and he had power here, but when the church rebel against him, and he had a few of the town folk went with him.

Have you paid attention to some of the symbols her" and he said to him," the ones that look like a Raven," with concern, and he said yes, well that the symbol of the Society, and at one-time people feared them?

It is said in time past, they use to sacrifice others that drifters in the town., in which I think they are out of mine, with a smile on his face, so peoples don't ask question.

The Agent asks him, does he have any family left here, and, 'is there someone else I can talk to about what in the hell is going on,

we have lost officers in these woods in the last few days, Sheriff, with a loud voice, while standing up, and said," look here Sir, when other people's start to come around here, this is what happens, we don't know a clue where they went.

After tension went down, the agent show him the picture he had in his possession, and ask him," is anyone here you know" and the Sheriff said," these two are dead, but I don't even know who the other person is, but if you go to the town Historian, he may help you, and he left.

Meanwhile in the woods, Raven was waiting for them to come back. Then he got a call and he said, 'listing, we must clean up this mess once and for all, you all know what to do, I got unfinished business here, I will contact you when it finishes, and by the way, 'kill them all, and he hung up the phone.

After hearing the loud conversation, they went back, and Stephen said," are you ready to go hunting", and she said," I thought you never ask."

About that time, Catrina stood down the hill for Raven to see here and she said to him, while having a spear in her hand, 'No more running" while frost was coming out of her mouth.

All of a curtain came out of the cave, looking around to see where the voice came from and she said, 'you fool, it's time to hunt, and before he could raise his raffle you to shoot, she took out with haste.

16

With a loud voice, he said to her," today you all will die" and he took off after them.

The Agent went into the courthouse to talk to the Historian, and he came up to his office and said," Sir I got some question for you, and I need some answer with a stern look on his face, and he told him to sit down.

The agent told the historian," do you know about the' 'Society", then suddenly, the historian said," you need the leave this place, before you and the rest of your officer get killed.

The agents were very shock of what he said and said to him," you don't have to be afraid, I know there are things that have been on in this town, and the woods relates to missing and the killing of other.

Then he told the Agents, with a fright look on his face, 'have anyone follow you over here, and the Agent said, 'no, why" and he told him," this town has secrets also, some of the people here still believe in the minister teaching, and peoples just don't talk, but they just do.

So, the Agents took the picture out of his pocket, and show it to the Historian and said, 'I know this is the late minister, and these two are dead, and the man in the woods, is Raven, but who is this one here.

About that time when he was about to tell him, there was a shot from a high powered rifle, and he feels dead, and a truck speed off.

The agents took cover and call his agents who were at the edges of the woods. The agent's phone rang, but there was no answer.

In his mind, he does know by now, the woods are the breeding fields for death.

Catrina was running down the muddy path, and while Raven was catching up, and about the time, she fell on the wood floor and when Raven got in range to take the shoot, he fell in the trap that Stephen had set for him and he knocked himself unconscious.

Stephen and Catrina walk toward the hole very slowly, pick up the gun, and they saw Raven, laying there, with no movement.

Then Stephen pulls him out of the hole, drug him back to the cave, and tired him up.

When Raven came to, he realized that he was tired up. He glazed around and said to them," what do you think this will do for your cause"

You all don't understand the society, we will never die, looking directly into their eyes.

Then Stephen said," for years your torcher me, and I have seen you not only kill peoples, but you also eat them.

You have taken my humanity out of me, and you hurt my love in front of my face.

So, I decided to grant you a slow and painful death. About that time, here comes Catrina, looking like a wild woman, she put the shock collar around his neck and turn the power up high as it can go, and made him pass out.

They both were sitting down, looking at Raven, and he said to her," baby I can't go back to civilization, because of the

events took here in these woods, but I
want you to go back,

you know why I can't go back with you
baby, with tears in his eyes," she said,
'baby I don't care what you did, and who
you became, but what I do know, I love
you" and she huge him with fear.

The agents don't know what to do, but he
knew, he had to get out of that town
soon.

So, he made his way to the police dept
and told the Chief, I need to call the
F.B.I. and the Chief said," all the tower
are down Sir', but we can help you.

The agents told them, we must get to the woods now, we need to warn them that we are dealing with a cult in the woods, with an exhausted voice, and they left with haste.

So, about that time, while they were driving the road, the agents said," is there a toward by for me to get some connection, and the Chief said," one mile before we get to your agents, you should be able to make the call to the Agency.

Stephen said to her," I know a way out Catrina, but we will have to leave soon, today is the first moon, and they will be here for worship, but before you leave, I must do this, so you would know this man is no more.

He set Raven upward, and start saying difference words, and after that, he looks over at Catrina and said, 'his power will be mind, and slit his neck from one end to other.

When the blood starts powering, he put a cup and caught about a cup full and drink it until it was no more.

She was afraid of what took place, but now she knows the man she craves for, is no more, and said," this is not your fault, but this demon here, but I want the man inside to know I will never stop loving him, with tears in her eyes and then he took her by the hand and left.

It was late evening when they got to the forest, and he saw the two cars that the agents were driving.

He jumped out of the Chief car with his gun in his hand, and the Chief got out also.

There was no moment in the cars, so the Agent told the Chief, "you go that way" and he did. About that time, he looked over in the car, the two officers were shot to death, and when he got over to the next car, where the other two was, one was shot in the head, and the other was missing.

Bout that time, the agent was holding his head down, the things started to get clear. When was asking everybody in town about the missing and killing, some of them had a mark on their middle finger, like the symbols around town.

While his head was still down, he remembers, when he was eating at the dinner, the symbols were there, and when the Sheriff stood up, he had the mark too.

Then he remembered, Raven last name was Mullins, and the person that was in the picture, was the Chief, his little brother and about that time, with his hand on the trigger he swirls and before he got the shot off, the Chief shot him in the head.

Later that even, when they came to the edge of the patch, she finally saw the road she needed to get home.

She said to him,' do you remember our first kiss Stephen" and he said, "until the day I die love", and she stood there, her face was pale, her hair was pull back, and look heartbroken.

He said to her, tell my sons, that I was found dead, and I want you to leave here, and never come back.

She took twenty paces out of the woods, and he was gone, and she left and went home.

Three days past and Catrina had packed up and got into u haul and pull off. While she was driving down the road, she began to think about the time, last time they made love.

Stephen came into the house on a long day of work, and when he opened the door, there was a letter on the coffee table said " I want to play a game with you tonight, but first, go to the bathroom, and get into the hot bath I made for you, and he did.

When he finished up, he looked over to his right, while the soap suds just pouring down like rain, there was another letter that said, when you get out, come into the bedroom, but don't cut on the light.

When he entered, with only a towel on, he heard voice said," Stephen I have been waiting on you for a while, with a passion in her voice,

Then she told him," you can turn on the light now, and what he saw, she was laying in the bed, with only stocking on, and a seductive look on her face.

He easy into the bed, spread her legs wide open, and her eyes went back into her head.

He was stroking her body so deeply, until about that time, here she comes, then from the blow of a car horn, she came back to reality.

She pulls over to the woods for the very last time, all the crime scene was going, so she blew her horn twice, and waited for him to show, but he did not come, so she started to pull off, and all a curtain, this figure appeared deep in the woods, and she smiles, and he turns around deep in the woods he went. Then she pulled off with a smile and was never seen anymore.

Later that night, you could see a fire deep in the woods, and the road at the entrance was full of cars.

As the crowd gathers in the circle, chanting world from a book in their hand, here comes a figure in a black hood, and it was Stephen.

When he killed Raven, he gained his strength and power, in his mind, The Chief pulls off his hood, and said," tonight we will feast off warm blood, and about that time, they brought out the agent that they did not kill.

They put him on a alter, while his tongue was cut out, and the Sheriff, took his hood off, and in his hand was a dagger, then he shouted out some words from an old book, and put the dagger into his chest, and death took the agents.

Two years have passed, and the woods keep its secrets, and Cartina has put the memories to rest.

She doesn't have to passion through anymore, but every now and then, she misses Stephen.

The next day near the park, a family that was headed on vacation, car broke down at the entrance of the woods, and the husband said," you all stay here, I am going to find help, The End.

34

www.ingramcontent.com/pod-product-compliance
Lightning Source LLC
Chambersburg PA
CBHW061236180526
45170CB00003B/1323